FORECASTING THE IMPACT OF TECHNOLOGICAL CHANGE ON MANPOWER UTILIZATION AND DISPLACEMENT: AN ANALYTIC SUMMARY

by

Alan Fechter

1215-1

March 1975

THE URBAN INSTITUTE
WASHINGTON, D.C.

The research and studies forming the basis for this report were conducted pursuant to a contract with the National Science Foundation, Office of National R & D Assessment. The statements and conclusions contained herein do not necessarily reflect the views of the U.S. Government or the National Science Foundation, which make no warranty and assume no responsibility for the accuracy or completeness of the information herein.

UI 1215-1
ISBN 87766-138-3

REFER TO URI 99000 WHEN ORDERING

Available from:

Publications Office
The Urban Institute
2100 M Street, N.W.
Washington, D. C. 20037

List price $2.50

A/75/750

TABLE OF CONTENTS

FOREWORD

This paper was prepared under the auspices of a National Science Foundation grant awarded by the Office of National R&D Assessment. It represents a six-month effort to survey and assess the literature on forecasting models and their ability to assess the impact of technology on manpower utilization. The views expressed in this paper are those of the author and do not necessarily reflect either those of the National Science Foundation or the Urban Institute.

The research effort was undertaken in collaboration with a research team from the University of Utah, which surveyed and assessed the literature on labor market adjustment mechanisms to the impact of technological change. The results of the two surveys are summarized and synthesized in a paper prepared by Garth Mangum, who was overall coordinator of the project.

The Urban Institute has also produced a companion volume, an annotated bibliography on manpower forecasting models, prepared by Pat Barry and Valerie James, which appears as a separately bound paper.

I am indebted to Rolf Piekarz and Barbara Burns, from the NSF National R&D Office, for their patience in dealing with the administrative problems involved in completing this study and for their reactions to earlier drafts of this paper. I would also like to express my particular gratitude to Pat Barry, Valerie James, and Julie Casamajor for their help in obtaining necessary library materials and my general appreciation of the library staff of the Urban Institute in performing above and beyond the call of duty in obtaining hard-to-get reference materials. Last, but not least, I want to thank Pat Coleman, Brenda Chapman, and Melissa Penney for their cheerful and skillful rendition of many drafts of my abominable handwriting into neatly typed manuscripts.

ABSTRACT

This paper contains an analytic summary of the state-of-the-art on technology and skill utilization. Specifically, it assesses the existing capability to forecast shifts in employment opportunities brought about by the diffusion of new technology. It is part of a larger effort investigating the socio-economic implications of technology.

Obstacles to producing forecasts of the impact of technological change on skill utilization are briefly discussed. These obstacles include the inability to forecast innovation, a theory of production that does not permit disaggregation of labor, an inability to isolate employment changes attributable to technology from employment changes caused by other factors and an unsatisfactory method of classifying skills.

Existing models used in forecasting manpower requirements are described and analyzed. These models are alike in that they assume that future requirements are determined by future output, future labor productivity, and future skill-mix. Two basic forecasting methods are described: inverted production function, and input-output analysis. Several types of manpower forecasts are discussed: long range estimates at the national level classified by skill(s), and regional models which are further disaggregated by skill. These types of forecasts are usually generated to assist in vocational and educational planning or in developing general manpower strategies for training or regional development. In addition, these forecasts are utilized by policy makers to evaluate the manpower consequences of alternative policy options. The survey revealed a concentration of models that produced long-range national forecasts and a dearth of models that generate short-range forecasts or forecasts disaggregated to the regional, state and local level. The models should have been evaluated on the basis of the accuracy of the predictions they generate compared to actual experience and/or the accuracy of the assumed changes in the independent variables on which the predictions are based. Since there is not much evidence on the accuracy of predictions, attention was focused on the reasonableness of the model structure. Most forecasting models are based on the dubious assumption either that relative prices and relative wages will remain constant or that output can only be produced with a fixed combination of inputs (i.e., there is no possibility of substitution). Moreover, projected changes in employment-related factors, such as output, productivity, and skill-mix are usually based on an assumption of either no change or changes based on past trends, thus ignoring cyclical effects or irregular changes in technology.

The models are also evaluated on the basis of their potential value in policy formulation. This value is assumed to depend on the nature of supply adjustments in the labor market. A review of the literature on supply behavior reveals that the evidence on adjustment is mixed. A considerable amount of mobility takes place, but supply elasticities are generally low. This suggests that, while adjustments do occur to shifts in labor demand schedules, they generally take the form of shifts in, rather than movements along, supply schedules. The implication is that forecasting models would be most useful in those markets in which mobility is expected to be relatively limited. Such markets include skills that are highly specific or that require a considerable amount of lead time for training.

I conclude that if resources are to be allocated to improving the state-of-the-art in manpower forecasting, they should be devoted to overcoming the obstacles discussed above and modifying the existing models so that they are more reasonable representations of labor markets. This would require more research on the determinants of such key forecasting variables as labor productivity and skill-mix.

I

INTRODUCTION

Technological change pervades and influences all segments of society--families, unions, businesses, schools, and government. The influence it has on employment has been a particular concern of policy makers for a considerable period of time. This concern reflects the disruption that is thought to occur in labor markets and its consequent cost, in terms of real output or human suffering. Sufficient warning of an imminent innovation, it is argued, would enable concerned policy makers to take appropriate measures to help ease the impending transition.

In the late 1950s, debate raged over the nature and causes of the persistently high levels of unemployment which characterized that period.[1] A National Commission on Technology, Automation, and Economic Progress, established by President Johnson in 1964, studied the role of technological change in the American economy and recommended relevant administration and legislative steps to cushion the employment effects of the adoption of new techniques.[2] Currently, the National Commission on Productivity, charged with the task of developing new programs and policies to improve productivity, has been concerned about procedures for the implementation of new technology.

New technology simultaneously makes some skills obsolete and generates demand for new skills. The resultant shift in employment opportunities concerns manpower specialists at all levels of government--federal, state and

1. See, for example, R. Solow.
2. National Commission on Technology, Automation and Economic Progress.

local. It is also of interest to a wide variety of other groups: education planners, employment counselors, vocational guidance specialists, personnel managers, and union officials. Of course, those most concerned are the workers whose skills are no longer needed.

Those responsible for assuring that labor market adjustments to new technology occur with a minimum amount of disruption and hardship might be aided in their task if they were able to anticipate these adjustments. This study evaluates the existing capability to anticipate these adjustments by means of forecasting models. Specifically, it attempts to assess the ability to forecast shifts in employment opportunities brought about by the diffusion of new technology. It is part of a larger effort investigating the socio-economic implications of technology. A companion study examines the state of our knowledge on how unions and businesses adapt to technological change.

Evaluation of our ability to forecast the implications of technological change on skill utilization depends on: (1) how well we are able to forecast future technological change; (2) how well we are able to translate this change into changes in employment and skill utilization; and (3) how important it is to be able to do this. The first issue is being addressed in other research efforts and lies beyond the scope of this study. It is therefore covered in only the most cursory fashion.

The second issue constitutes the heart of this paper. Forecasts of manpower requirements (employment) and skill utilization are usually predicated on projections of output, labor productivity, and skill mix within an activity. Sometimes the assumptions underlying these projections are explicit; frequently they are not. In all cases, technology is but one of many factors operating on these variables.

The third issue relates to questions of adaptability in labor markets and the possible role of government policy. If labor markets were highly flexible and adaptable, particularly on the supply side, there would be little reason to forecast the impact of technology on labor markets since these anticipated changes would not be very costly, either in terms of real output or in terms of the hardships that fall on displaced workers and their dependents. On the other hand, labor markets in which the supply is highly inflexible, either because of obsolescent skill or because of prohibitive costs involved in adaptation, are targets for various forms of government intervention either to reduce these costs or, in some other way, to alleviate the adverse consequences of this inflexibility.

The above considerations move us to conclude that the best way to evaluate models forecasting the impact of technology on employment and skill utilization is first to evaluate more general models of manpower requirements, focusing on the role technology plays in these models, and then to assess adaptability on the supply side of labor markets.

There are many possible ways to define technological change. For this particular study, an appropriate way to view it is to define it in terms of the parameters of the production function; i.e., technological change can be defined as something that changes the amount of output that can be produced by a given combination of inputs, or the rate at which inputs can be substituted for one another in order to produce a given level of output. As we shall see, such a definition of technology is difficult to quantify. However, observable implications of technological change would include changes in factor productivity and changes in skill-mix. Unfortunately, factor productivity and skill-mix are also affected by variables other than technology. Thus, observed changes in either cannot necessarily be solely attributed to technology.

In the following section some obstacles to producing forecasts of the impact of technological change on skill utilization are briefly discussed. These obstacles include the inability to forecast innovation, a theory of production that does not permit disaggregation of labor, and inability to isolate employment changes attributable to technology from employment changes caused by other factors and an unsatisfactory method of classifying skills.

In Section III, existing models used in forecasting manpower requirements are described and analyzed. The models are evaluated in Section IV on the basis of the accuracy of the predictions they generate compared to actual experience and/or the accuracy of the assumed changes in the independent variables on which the predictions are based. Since there is not much evidence on the accuracy of predictions, attention is focused on the reasonableness of the model structure. Most forecasting models are based on the dubious assumption either that relative prices and relative wages will remain constant or that output can only be produced with a fixed combination of inputs (i.e., there is no possibility of substitution). Moreover, projected changes in employment-related factors, such as output, productivity, and skill-mix are usually based on an assumption of either no change or changes based on past trends, thus ignoring cyclical effects or irregular changes in technology.

In Section V, the models are evaluated on the basis of an additional criterion: their potential value in policy formulation. This value is assumed to depend on the nature of supply adjustments in the labor market. A review of the literature on supply behavior reveals that the evidence on adjustment is mixed. A considerable amount of mobility takes place, but supply elasticities are generally low. This suggests that, while supply adjustments do occur, they generally are in response to changes in non-wage supply determinants. The implication is that forecasting models would be most useful in those markets in which mobility is expected to be relatively limited. Such markets include skills that are highly specific or that require a considerable amount of lead time for training.

In Section VI, the current state-of-the-art is summarized and future directions for research are suggested. I conclude that, if resources are allocated to improving the state-of-the-art in manpower forecasting, then they should be devoted to overcoming the obstacles discussed above and modifying the existing models so that they are more reasonable representations of labor markets.

II

OBSTACLES TO FORECASTING MANPOWER REQUIREMENTS

The capability to forecast the impact of future changes in technology on skill utilization can be conveniently broken down into two steps: (1) the forecasting of future technology, and (2) the assessment of its impact on the demand for labor, disaggregated by skill. The former step involves understanding the factors that influence research and development on new technology, the factors that determine what developments get innovated into the production process, and the rate of diffusion of this innovation. The latter step requires the isolation of the partial impact of technological change on the demand for labor, disaggregated by skill. The partial impact can, in turn, be decomposed into two further components: that which, *ceteris paribus*, changes the relative demand for different factors of production by altering the relative factor cost of production, and that which, *ceteris paribus*, changes the demand for all factors of production by altering the profitability of production. These components are frequently discussed in terms of the "neutrality" of technological change.

Studies of past experience indicate that, while technological change has been and can be expected to continue to be a major cause of economic growth, we are a long way from understanding its determinants.[1] Factors such as firm size and market structure do not appear to be important in producing technological progress; indeed, there is speculation that these factors work in offsetting ways upon research and development and innovation,

1. See Kennedy-Thirwall, pp. 44-62.

the two major components of technical change. Moreover, the existing studies show that, even if we had some conception of what the major determining factors were, the amount of time it takes to transform a development into a new innovation varies greatly. Given a paucity of empirical evidence about what determines the development and diffusion of new technology, it is small wonder that our capability to forecast technological change is limited. Most technology forecasting is predicated upon highly subjective methods, such as Delphi techniques, forecaster judgment or trend extrapolation. Needless to say, there is considerable room for improvement in this area. Hopefully, new research will produce a firmer basis for such forecasts in the future. For now, available knowledge limits us to tolerable forecasts of changes that have passed the development stage and are about to become innovations.[1] Moreover, this limited capability to forecast technological change does not allow us to accurately forecast its rate of diffusion in specific industries because this rate has been extremely low on average and highly variable among industries.

Assessment of the impact of new technology on the demand for labor, disaggregated by skill, is limited by existing production theory and by inadequate treatment of the concept of skill. In order to isolate the partial impact of technology on the demand for labor of differing skill capacities we must first be able to specify a production function that relates ouput to inputs in which labor is disaggregated by skill. Traditional production functions limit themselves to capital and labor in the aggregate. Attempts are being made to modify this theory to allow for further disaggregation,[2] but these modifications have not yet been

1. Haase, p. 63.
2. An example of this new work may be found in Christensen-Jorgensen-Lau.

empirically treated in any extensive way. Past restriction to two-factor production functions has limited the capability to identify past changes in the demand for labor arising from technological change to aggregate employment. Moreover, even in the case of two-factor production functions, it has been difficult to disentangle changes in the demand for labor resulting from other factors, such as changes in the relative cost of labor.

In principle, technological change can affect the demand for labor in offsetting ways. A technological change, by raising factor productivity, can reduce the demand for labor for a given level of output. On the other hand, by making production more efficient, technological change increases the profitability of producing a given level of output and provides the incentive to increase output and, therefore, to increase the demand for labor. Available evidence is consistent with the conclusion that technological change has expanded the demand for labor in the aggregate. Existing studies show that growth in output has been the result of growth in inputs of production and growth in total factor productivity.[1] The latter represents an index of technological change.

While technological change appears to have stimulated aggregate employment, its impact on labor disaggregated by skill has been less thoroughly documented. As noted above, a major obstacle in studying the effect of technology on skill has been the restriction to a two-factor production function.

---

1. Most of the studies have attributed 80 to 90 percent of the long-run growth in output to growth in total factor productivity and 10 to 20 percent to growth in factor inputs. However, these studies have treated factor inputs as homogeneous units, not subject to quality change. Adjustments for such quality changes would probably reduce the percent of long-run growth in output attributable to growth in total factor productivity and increase the percent attributable to growth in inputs (adjusted for quality). For survey of the literature on the output effects of changes in technology, see Kennedy-Thirwall, pp. 13-43.

In addition to being able to anticipate technology and specify an appropriate production function, skill must be definable in operationally meaningful ways. Ideally, we would like a classification scheme that enables us to group our employment statistics so that the elasticity of substitution in production and the elasticity of skill supply are relatively high within classes and relatively low between classes.[1] Unfortunately, there is little evidence available on these elasticities. Since existing statistical dimensions of skill, such as Census definitions of occupation, are thought by many to be inadequate, experiments are being made with alternative classification schemes.[2]

All studies reviewed for this survey, regardless of their definition of skill, show trends toward higher skill levels which are projected to continue into the future. Unfortunately, there is little evidence about the determinants of this trend. It could reflect technology, changing relative factor costs, or some combination of these.

1. See Welch for an interesting discussion of this problem. Also, see Cain-Hansen.

2. Scoville constructs a set of job families and a set of job-content levels within each job family based on the structure of the job and the skills required to perform the job. Wool constructs an occupational index based on the socio-demographic characteristics of those employed in the occupations. Both are essentially alternative aggregations of Census occupations. See Scoville, pp. 11-33; Wool, pp. 31-45. See also Bezdek (1973), pp. 47-53, for a summary and critique of existing occupational classification schemes.

III

EXISTING MANPOWER FORECASTING MODELS

Since the objective of this paper is to evaluate models forecasting the impact of technology on manpower utilization, I limit my survey to models of manpower requirements. There is an analytic framework that is implicit in all of the work I reviewed. It is a model of employment as a derived demand, and it can be summarized as follows:

$$(1)\quad L_{si} = \alpha_{si}\beta_i X_i$$

where L represents employment, X_i represents output, β_i represents the reciprocal of average labor productivity (L_i/X_i), and α_{si} represents the share of employment for a particular skill (L_{si}/L_i). The subscripts s and i stand for the particular skill and industry for which a forecast is being made. Differences among forecasting models can essentially be identified in terms of differences in their assumptions about β_i and α_{si}. Most models assume that they are fixed values, not subject to change, or that they change in mysterious ways that can be projected from past trends or benchmark data.

Given equation (1), predictions of future employment can be generated on the basis of forecasts of future changes in output, labor productivity, and skill-mix:

$$(2)\quad \frac{\partial L_{si}}{\partial_t} = a_1 \frac{\partial \alpha_{si}}{\partial_t} + a_2 \frac{\partial \beta_i}{\partial_t} + a_3 \frac{\partial X_i}{\partial_t},$$

where $a_1 = \beta_i X_i$, $a_2 = \alpha_{si} X_i$, and $a_3 = \alpha_{si}\beta_i$. Several interesting features emerge from this formulation. First, accuracy in the employment projections requires accuracy in the forecasts of future changes in output, labor productivity, and skill-mix. Second, and less obvious, the three factors are erroneously assumed to be independent of each other. The assumed employment impact of a change in labor productivity, for example, will depend on the value of a_2. This value, in turn, is determined by the product of the skill-mix and output. Assuming that a_2 will be stable implies either that there will be no future changes in output and skill-mix or, if such changes are anticipated, that they will be offsetting in their impact on a_2.

In the following review of manpower forecasting models, I first describe basic methods employed in generating manpower forecasts. I then describe the types of manpower forecasts that have been generated. I close with a short summary of the gaps in the existing literature and possible reasons for their existence.

Existing Models

Forecasting models may be classified into two generic techniques: inverted production functions and input-output. The inverted production function technique generally involves projecting future employment from the inverse of a production function.

$$(3) \quad L = f^{-1}(X, K)$$

where f^{-1} is the inverse of the production function relating L and K to X, holding technology constant. These models are usually used to generate forecasts of total employment (either in the aggregate or by industry) and forecasts of skill-specific employment (particularly for occupations). Projected employment is generated from projected values of varying combinations of X, K, and technology.[1] For forecasts of total employment, econometric models to estimate the parameters of the inverse function generally regress L against estimates of varying combinations of X, K, and a trend variable. The latter variable is included to capture the effect of technological change (if K is also included) or the joint effects of growth in the capital stock and technological change (if K is excluded). Because production functions are usually two-factor equations, employment estimates generated by this method cannot be disaggregated by skill within industries. For skill-specific employment, forecasts are usually generated from projections of output (X) and assumptions about the ratio of employment to output. These assumptions take the form of either a positive projection of past trends or a normative assertion of what a desirable ratio should be.[2]

The input-output technique is a special case of inverted production functions that allows for the effects on employment of the existence of intermediate products; i.e., outputs of one industry that are used as inputs to another industry. Thus, a given change in final demand can change the amount of output demanded directly, through the change in output in the industrial sector required to satisfy that demand, and indirectly, through the change in outputs in industrial sectors that produce inputs to that particular sector. The procedure may be summarized in three steps: the conversion of final demand into output by industrial sector; the translation of that output into employment by industrial sector; and, on occasion, the allocation of that employment by skill.[3]

---

1. For example, see Scott, McCracken.
2. See, for example, National Education Association, Altman, Cartter.
3. See Almon, Bezdek (1972), Lecht, BLS (1966, 1970, 1972a, 1972b) for illustrations of this method.

Conversion of final demand to output by industrial sector is accomplished by means of a table of input-output coefficients. To illustrate, let X be a vector of total outputs by industry, Y be a vector of final demands, T be a matrix of intermediate products, and A be a matrix of input-output coefficients measuring the inter-industry flow of intermediate product per unit output. The traditional definition of GNP can then be written:

$$(5) \quad Y = X - T$$

T, in turn, can be expressed in terms of X through the coefficients matrix if intermediate products are used in fixed proportions to output:

$$(6) \quad T = AX.$$

Substituting (6) into (5) and solving for X produces:

$$(7) \quad X = (I-A)^{-1} Y,$$

where I is an identity matrix. From (7) the output of any industrial sector can be linked to final demand. Several models have been developed to translate estimates of macroeconomic variables into industry detail. These models attempt to provide a set of estimates of industry output or sales that are consistent with the relationships between GNP components and industry output implied by the input-output tables.[1] These models include the Brookings Model,[2] the Wharton Annual Model,[3] the IBM Model,[4] the Sundarajan Model,[5] the Maryland Model,[6] the Economic Growth Project Model,[7] and the Illinois

1. See Bezdek (1973), pp. 9-31 for a detailed account of the alternative methods used by different models to accomplish this linkage.
2. Fisher, et al, Klein - Fromm; and Kresge.
3. Preston.
4. Bezdek (1973), pp. 19-21.
5. Sundarajan (1970, 1971).
6. Almon (1970a, 1970b).
7. BLS (1966b, 1970, 1972b).

Model.[1] The input-output coefficients, while assumed to be fixed in any given year, are allowed to vary over time. Forecasts of future values of these coefficients are based on estimated trends and judgment.[2]

Outputs by industry are converted into estimates of employment by industry through projections of labor productivity by sector. These projections are generated on the basis of equations estimated as part of simultaneous systems of equations describing the entire economy,[3] ordinary least-squares regression analyses[4] or some other extrapolation procedures.[5]

Allocation of employment within industry by skill, when attempted, is generally accomplished by projecting past changes in within-industry skill-mix. These past changes are usually derived from decennial Census reports.[6]

Types of Forecasts

Existing manpower forecasting models have been motivated by a variety of objectives.[7] A primary objective is educational and vocational planning. Manpower forecasts are needed to estimate future skill imbalances so that education and manpower specialists can reallocate resources away from the anticipated areas of surplus and to the anticipated areas of shortage. The type of model required generally varies according to the type of planning being done.

---

1. Bezdek-Getzel, Bezdek-Hannon, Bezdek (1973), pp. 29-31.
2. See BLS (1970), pp. 65-72, for a discussion of one revision process.
3. The Canadian manpower analysts are beginning to explore this option by integrating elements of their macroeconomic CANDIDE model into their projections model. I am indebted to Jim Hamilton of the Canadian Department of Manpower Immigration for his help in acquainting me with the activities of Canadian manpower forecasters.
4. See, for example, BLS (1966, 1970, 1972a, 1972b).
5. In those cases where productivity had to be projected on some other basis, the procedure varied according to how much the projected growth rate varied from historical growth rates. BLS (1970) also uses this technique for some industries.
6. See, for example, Scoville, p. 53; Wool, pp. 248-251.
7. See Mangum-Nemore, pp. 7-11, for a discussion of these.

Educational planning generally requires models that produce long-range forecasts of manpower classified by level of schooling completed or for very specific occupations (usually those requiring post-graduate training, such as medical doctors or scientists with advanced degrees) or highly specific college training, such as teachers, engineers, or nurses. Vocational planning generally requires models that produce intermediate or short-range forecasts of manpower classified by occupation and/or industry.

An additional objective to be served by manpower forecasting models is that of more general policy planning. A particular example of such planning is the use of long-range manpower forecasting models to check the consistency of overall growth objectives with their manpower implications. For example, what would be the manpower implications of a growth policy that will produce a given rate of unemployment--say, four percent--by a given year--say, 1980? This type of question is asked by both national and local policy planners. Or, what would be the manpower implications of a major policy change, such as a reduction of the Federal budget for defense spending, an increase in the Federal budget for domestic programs, or the adoption of a major revenue-sharing program? Frequently, this type of objective requires models that produce intermediate or long-range forecasts of manpower classified by occupation and/or industry and/or region.

Because of these differences in objectives, existing manpower forecasts can be classified into three types: (1) national forecasts which are further disaggregated into region and/or skill (i.e., education, industry, and/or occupation); (2) national forecasts for specific skills; and (3) regional forecasts which are further disaggregated by skill.

National Forecasts

National forecasts which are further disaggregated into regions have been generated by the National Planning Association (NPA) models for the United States, and the Ahamad model for Canada.[1] Two NPA models were surveyed: one which projects regional estimates of total employment, and one which projects regional estimates of employment by industry and occupation. The former model combines trend extrapolations for "basic" industries within a region with an "export-base" model of the regional economy to derive estimates of total employment by region.[2] The latter model uses essentially the input-output method to derive national estimates of employment by industry and occupation and allocates these national totals to regions by means of a "differential and proportional shift" analysis.[3] This shift analysis takes into account the industrial composition and the competitive position of the region in generating employment forecasts.

The Ahamad model for Canada derived employment estimates at the national level from independent projections of output and labor productivity. Employment by industry and region were forecast by means of trend extrapolation. Distributions by occupations within industries were also forecast at the national level on the basis of trend extrapolation. These distributions were generated on a regional level by assuming that the regional occupation structure, derived from 1961 Census data, would change at the same rate as the national structure.

---

1. Ahamad.
2. Scott.
3. Darmstadter.

National forecasts which are further disaggregated into occupations and industries include those generated by the BLS and the NPA models. The BLS model is perhaps the best known of all the forecasting models. Initiated in the mid-1960s, the BLS model was first used to generate projections of employment by occupation and industry for the year 1970 for the National Commission on Technology, Automation and Economic Progress.[1] Since that time, it has also developed employment projections for the years 1975, 1980 and 1985.[2] Aspects of the BLS model have been used by a number of other studies.[3]

Both models utilize the input-output method. GNP is projected on the basis of an assumed rate of unemployment. This is translated into a vector of outputs by industry on the basis of projected values of the parameters of the input-output matrix. This vector, in turn, is converted into a vector of employment by industry by means of projected values of the inverse of labor productivity. Finally distributions of occupations within industry are projected on the basis of past trends in Census data.

Specific Skill Forecasts

National forecasts for specific skills are generally dominated by models for forecasting particular professional skills, such as teachers, scientists, engineers, nurses and medical doctors. Forecasts are usually generated by means of an inverted production function which relates

1. BLS (1966).
2. BLS (1970, 1972a, 1972b). Alterman, Rosenthal.
3. Wool, pp. 218-260.

employment to some measure of, or proxy for, output. Cartter, in his study of college teachers, projects demand from independent estimates of future enrollments. Similar techniques have been used in studies made by the National Education Association in the United States and the National Advisory Council on the Training and Supply of Teachers in England. Altman has projected employment of nurses on the basis of population.[1] Employment of scientists and engineers has been projected in a variety of studies on the basis of total employment in particular activities and the share of that employment that belongs to the particular scientists and engineers being analyzed.[2] Some of these studies also project the replacement demand for persons who leave the occupation or skill for employment elsewhere, retirement, or death.

Regional Forecasts

Forecasts of employment for specific regions are generally based on input-output methods. Such forecasts have been generated for Hawaii,[3] the District of Columbia,[4] Denver,[5] the state of New York,[6] Oregon,[7] and Kansas.[8]

Summary

Two types of manpower forecasting methods were discussed: those based on production functions and those based in input-output analysis. These

---

1. Altman (1971), pp. 152-156.
2. See, for example, Fortune-Ross; Ross; National Science Foundation. See also Bain, pp. 99-124 for a survey of the forecasting literature for scientists and engineers.
3. Ferber-Sasaki.
4. Manpower Administration, District of Columbia.
5. McCormick-Franks.
6. National Planning Association.
7. Watson.
8. Spellman.

methods have been used to developed a variety of manpower forecasts at different levels of aggregation.

The specific manpower forecasts that have been developed were generated in response to specific planning objectives of particular groups of people. Vocational guidance counsellors and education planners usually require long-range forecasts of manpower by skill at the national level. Planners associated with manpower training programs usually require intermediate-range forecasts of manpower requirements by skill at the regional level. Officials responsible for health or educational planning usually require intermediate- and long-range forecasts of manpower in specific skills (e.g., doctors, nurses, teachers, etc.) at both the national and the regional level.

Existing forecasts reflect the historical development of these needs. Until the early part of the 1960s, the predominant demand for manpower forecasts came from the vocational guidance counsellors and the education planners. These people needed mainly long-range skill forecasts at the national level. The early sixties saw the growth of regional development and manpower training programs. And these programs generated new requirements for shorter range forecasts at a lower level of regional aggregation.

As a result of this history, models that produce long-range forecasts of manpower skill requirements at the national level appear to be the most prevalent and, among these, models based on input-output analysis, are the most widely used. Recently, however, models that produce forecasts of manpower requirements at the regional level have begun to make an appearance. In part, the emergence of these models is due to the encouragement that has been provided by federal funding. On balance, however, there appears to be relatively few models that generate regional forecasts and few models that generate short-range forecasts of manpower requirements by skill. Possible reasons for these gaps include: (1) an inability to project accurately short-run changes in key variables of the forecasting models (e.g., labor productivity, output, and skill-mix), and (2) the lack of adequate regional data bases. These reasons will be discussed in more detail in the following section.

IV

EVALUATION OF MODELS

Discussions of model evaluation are not usually confined to forecasting models. Rather, they examine a wide range of models which, for ease in exposition, are classified into several general classes. Fromm, for example, discusses three types of analytic models--policy simulations, forecasting, and structural. The classes generally overlap in their use of methods and are mainly distinguishable on the basis of the objectives of the researcher.

A relatively clean method of classifying empirical models would be one which distinguishes "behavioral" models from "simulation" models. Behavioral models are defined as models whose major objective is to infer behavioral relationships from observed data. Most econometric models would fall into this class of models. Behavioral models are judged on the basis of how well they "explain" observed behavior, how well the relationships estimated from the behavior accord with a priori expectation, and how well they predict beyond the range of observed behavior. The reality of assumptions underlying the models is frequently questioned. Some argue that the reality of these assumptions is not at all relevant to evaluation of the worth of the model. Others, on the other hand, maintain that realistic assumptions must be made if a model is to perform adequately.[1]

1. For a discussion of this particular methodological question see Friedman, pp. 16-30; Koopmans, pp. 129-166 (see especially pp. 135-142).

Simulation models, on the other hand, have as their major objective the prediction of outcomes on the basis of a given set of behavioral relationships. They can be characterized as attempts to answer the question: What would happen if . . . Policy simulations, for example, ask what would happen if a particular policy, e.g., a reduction in defense expenditures, were adopted. These types of simulations are generally used to aid policy planners in choosing among alternative options. Such models can be judged on the basis of the accuracy of their predictions. These predictions, in turn, are dependent on the choice of variables and behavioral relationships used to construct the simulation model.

Forecasting models are forms of simulation models. They also ask the question: What will happen if. . . They attempt to predict future outcomes on the basis of a given set of behavioral relationships. Their structure can be described in terms of a set of independent variables and a set of relationships between these variables and the variable to be predicted. And they can be evaluated on the basis of the accuracy of their predictions and/or the validity of the variables and relationships chosen to constitute the model. Since forecasting models frequently differ in their objectives, the standards used in evaluating them cannot be uniform.

Ahamad and Blaug discuss three types of manpower forecasts--policy-conditional, onlooker, and optimizing. Policy-conditional forecasts assume values of some independent variables based on the achievement of a given policy-objective. Models which generate forecasts based on the assumption of full employment (a given rate of unemployment) fall into this class, and the NPA model and the BLS model used by the National Committee on Automation, Technology, and Economic Progress are examples of such models.[1] Many have been critical of these studies for not considering the implications of alternative rates of unemployment.[2] Onlooker forecasts are distinguished by assuming reasonable or most-likely values of the independent variables in contrast to values dictated by policy objectives. These values are chosen on the basis of past trends, judgment, or explicitly specified models. Optimizing forecasts assume values of independent variables which will maximize or minimize some objective function subject to some constraints.

One cannot compare models producing policy-conditional or optimizing forecasts with models producing onlooker forecasts. In the case of the former models, a key factor to consider is the validity of the model. In particular, one needs to ask whether it includes all the relevant independent variables and whether the assumed relationships between the independent variables and the variable to be forecast are accurate, stable, and/or reasonably well-predictable. In the case of the latter model, a key factor to consider is the accuracy of the projection. The structure of the model is of secondary importance. Methods of assessing the accuracy of forecasting models have been discussed in a number of studies.[3] Ideally, the method that has most appeal is one which allows the forecaster to assess the accuracy of his forecast <u>and</u> to modify the structure of his model accordingly.[4]

---

1. Lecht and BLS (1966).
2. See, for example, Striner; Hansen.
3. See, for examples, Stekler; Mincer-Zarnowitz; Theil (1964, 1966); Klein.
4. Mincer-Zarnowitz, pp. 6-12; Fechter, pp. 23-27.

Unfortunately, little effort has been made to evaluate the accuracy of the forecasting models reviewed or to compare the relative accuracy of alternative forecasting methods.[1] Among the reasons offered for not assessing the predictive ability of the models is that, frequently, events overtake the assumed conditions of the model. Thus, projected shortages do not materialize because actions are taken to alleviate them. Some analysts even go so far as to claim that the failure of a forecast to materialize is a measure of the success of the forecasting activity. In other words, if a forecast is made to project skill imbalances so that they can be alleviated, then the failure of a projected shortage to appear because of actions prompted by the forecast can really be counted as a successful application of the forecasting model. In essence, these analysts are advocating that forecasting models be judged by how well they achieve their policy objectives as opposed to how accurately they predict events. While there may be a legitimate argument against judging forecasting models solely in terms of their predictive accuracy, there is no legitimate excuse for failing to compare the performance of the model to actual experience--particularly in the case of continuous forecasting activity--if only to provide a basis for improving the structure and, therefore, the precision of the model.

In the absence of direct assessment and analysis of forecasting errors an alternative criterion, the reasonableness of the assumptions that underlie the forecasts, is examined.

---

1. Exceptions to this assertion include Swerdloff, who evaluates the accuracy of BLS projections for the 1960s and finds that aggregates are reasonably accurate, but disaggregates, notably women, older workers, and selected industries (service, construction, government, agriculture, and mining) are not. Ferber-Sasaki also evaluate the accuracy of their model in comparison to two more naive models and find that their model produces more accurate forecasts. For evaluation of the accuracy of input-output models, see Christ, pp. 158-168; McGilvray-Simpson, especially pp. 217-220.

Earlier, I noted that most forecasting models are specified, either explicitly or implicitly, in terms of the following variables: output, average labor productivity, and skill-mix. These are the key variables whose future movements must be projected in order to derive forecasts of employment by skill.

Ahamad and Blaug[1] note that few forecasts of the "demand" for manpower fit the economist's definition of demand as a function of price. More typically, forecasts of manpower "needs" or manpower "requirements" are made on the assumption that all relative prices, wages and salaries remain constant. The assumption of a fixed relationship between the outputs and the inputs of the input-output matrix implies either stable relative input prices or fixed input coefficients of production.[2] A similar interpretation can be made of an assumed stable average labor productivity. Labor productivity can be affected by relative factor proportions, technology, and possibly output. Relative factor proportions are, in turn, influenced by relative input prices. Stability in labor productivity implies either stability or offsetting movements in all of these factors. There is evidence that in input-output models, the labor coefficient is considerably less stable than the other interindustry coefficients.[3] Finally, a stable relative skill-mix is also consistent with a fixed coefficient production function or constant relative wages, no change in technology, and no change in output.

---

1. Ahamad-Blaug, pp. 8-9.
2. Bezdek (1973), pp. 32-39 describes possible factors influencing input-output coefficients and methods currently used to modify them.
3. Bezdek (1973), Table 2, and pp. 43-44.

Thus, following Hollister, I conclude that, for manpower projection purposes, a low elasticity of substitution is "desirable" because it would improve the forecaster's ability to predict skill requirements from a given level and composition of output. The reason for this would be that the observed skill composition on which forecasts are based would reflect technological conditions rather than relative supply conditions.[1] The evidence on the elasticity of substitution between labor and capital is substantial and the range of estimates varies from industry to industry and averages around one.[2] The evidence on the elasticity of substitution among skills is sparce and what there is produces a wide range of estimates.[3] This evidence implies that manpower forecasting models should change their focus from point estimates of future skill requirements based on an assumed fixed-proportions type of production function and begin to develop models that will produce a range of estimates of future skill requirements contingent upon the relative supply of skills that is projected to be available.

Even with the modification of existing models to allow for possible substitution between labor and capital and among skills, forecasting errors can be expected because we are unable to accurately project future values of output, productivity, and skill-mix. It would appear reasonable to conclude that projections of output will be more accurate in long-range models than ib short-range models, and in aggregate, as opposed to disaggregate models.[4] Part of the rationale underlying this conclusion is that larger

---

1. The observed skill coefficients would be insensitive to differences in relative supply (and resultant relative price differences).

2. See Nerlove for a review of this evidence.

3. Bowles; Psacharopoulos-Hinchliffe; Dougherty; Gramlich. The incidence cited in Bezdek (1973), pp. 43-44 is also relevant here.

4. Particular types of disaggregation include disaggregation by region industry, occupation or education.

proportions of the observed long-run variation in output is explained by factors with strong trend components. Short-range variations in output are more likely to reflect cyclical factors which are not adequately accounted for by current projection techniques.[1] Output measures are more difficult to project in disaggregate studies for a number of reasons. For regional models, one must typically assume something about regional shares of national output or, alternatively, in regional input-output models, one must assume something about the stability of the trade sector of the particular region being analyzed. In the case of the former, the assumption about regional shares is in addition to assumptions about national trends, providing opportunity for still larger errors in projections of regional output. In the case of the regional input-output model, the trade sector of a nation. is a larger share of the total economy than the trade sector of a nation. Thus, errors arising from violations of assumed future stability in the regional coefficient will create larger errors in regional forecasts.

For occupational or educational models, output is just more difficult to measure. Population is frequently used as a measure of output in projecting requirements for doctors when what is wanted is a measure of the expected number of people by type of illness or disease. Similarly, projected enrollments are frequently used as a proxy for the output of teachers when what is really wanted is a measure of the learning these students have undergone. Moreover, even if the outputs are measured correctly, many occupational projection models develop mechanical projections that are based on normative values of what the ratio of output to occupational employment should be.

1. For a suggested method of improving on these techniques, see Klotz.

Some analysts have questioned the judgment underlying the choice of these normative values.[1]

Future changes in labor productivity are generally projected using simple extrapolation or regression techniques or judgment.[2] These future rates of change are assumed to be stable. Unfortunately, there is an abundance of evidence that there is substantial variation in rates of change in labor-productivity, both within and among industries.[3] A review of some of this evidence has led one analyst to conclude that, ". . .it is clear that any attempt to deal with the productivity change problem in terms of simple extrapolations based on recent changes is likely to be very misleading."[4] This high degree of instability suggests that research aimed at explaining the determinants of this instability may improve our future capability to project changes in labor productivity.[5] This further suggests that even long-range aggregate forecasts of changes in labor-productivity may be susceptible to substantial error arising from such factors as unanticipated shifts in industrial composition or technology.

Future changes in the skill-mix, usually measured by occupational or educational distributions, are also generally projected by means of simple extrapolation techniques or judgments. Frequently, forecasting models are confined to particular occupations, like teachers or engineers.[6] These studies do not require estimates of skill-mix; they generate their skill forecasts on the basis of trend, or predicted change in output or population.

---

1. See, for example, Mangum and Nemore, p. 13.
2. See, for example, BLS (1970).
3. Kendrick; Massell; Fabricant.
4. Hollister, p. 388.
5. The work of Fair, Nordhaus, and Perry, among others, represents such research.
6. For examples, see Cartter; Engineering Manpower Commission; and National Education Association, Research Division.

But other studies attempt to be more comprehensive. Some of these start from estimates of total employment by industry and general projections of future skill-mix by industry on which they base their forecasts of future skill utilization.[1] One suspects that skill-mix is somewhat sensitive to the business cycle and, to the extent that this is true, short-run projections will be less accurate than long-range projections. In addition, skill-mix can be sensitive to technological change, if the change is not neutral. The continuing debate on the impact of automation, while focusing on the labor-capital tradeoff, also produces allegations that it is eliminating low-skill jobs. Existing evidence is inconclusive.[2]

To summarize, forecasting models are difficult to evaluate because of differences in their objectives and because few forecasters have put their models to the ultimate test: the comparison of actual to predicted outcomes. While such comparisons have not been made, they are feasible and could provide valuable information for both evaluation and subsequent refinement of these models. Resources ought to be devoted to such activity in the future. Given the lack of direct evidence on model reliability, I examined the reasonableness of the assumptions underlying the forecasting models. I found that models of manpower requirements are generally predicated on an assumed zero or low elasticity of substitution among skills and between labor and capital and/or an assumed fixed technology. Existing evidence on the elasticity of substitution indicates that it is not low and that, therefore, these models ought to be adapted so that they can produce

---

1. For examples, see BLS (1972a); Scoville.

2. Bright; Horowitz-Hernstadt. Horowitz-Hernstadt show that technological change does not systemtically favor high skills. On the other hand, Scoville and Wool forecast future declines in the employment of low-skilled labor. Since the latter forecasts reflect factors in addition to technological change, the two sets of findings are not necessarily inconsistent.

conditional estimates of future skill requirements based on alternative assumptions about relative skill supplies. Such modifications will weaken the policy making implications of these models, but, hopefully, they will also make them more accurate in their depiction of the future skill situation.

I also concluded that, even with more refined assumptions, our capability to generate accurate forecasts of future skill requirements is limited by our inability to project future values of output, productivity, and skill-mix. This inability reflects our ignorance about their determinants and can be overcome by more research to reduce this ignorance.

V

POLICY-RELEVANCE OF MODELS

This review of the state-of-the-art in forecasting the impact of technology on manpower utilization has thus far shown that: (1) it is extremely difficult to project the introduction of new technology to the production process; (2) it is also difficult to disentangle the manpower effects of technology from the manpower effects of the other factors, such as changing relative prices in product markets or factors markets; (3) existing production theory does not allow us to deal adequately with labor demand disaggregated by skill; (4) the assumption of limited substitutability among skills and between capital and labor are not borne out by the evidence and the models ought to be modified to reflect this evidence; and (5) even if the model was well-specified, the ability to accurately project key variables is quite limited. In short, the existing state-of-the-art can be classified as primitive. Much more research must be done on the determinants of innovation and productivity and the parameters of the production function before more accurate and refined forecasts can be generated.

The question remains, however, as to whether more accurate and refined forecasts are necessary. Such models would be necessary if our inability to accurately forecast the impact of technology on skill utilization resulted in substantial amounts of social cost, in the form of unemployment, inadequate wages and human suffering arising from the inability of the market to adapt to this impact. Such inabilities would be reflected by inelasticities on the supply side of the labor market; in particular, in the form of low short-run supply elasticities arising from the highly specific skill-content of the labor force or the relatively long training periods required to obtain the requisite skills. In this section, I review the literature on the supply side of the labor markets in order to evaluate their adaptability. I begin with models of relative supply and labor mobility. I conclude with a review of unemployment statistics and the possible location and magnitude of labor markets that might benefit from improved forecasting models.

Labor Force Participation

Labor force participation rates have always been an object of scholarly interest. Empirical analysis of this behavior dates back to the works of Woytinsky, Durand, and Long, but began in earnest after the famous contribution of Mincer.[1] In principle, we can separate these studies into two classes: time-series studies which focus on the impact of employment conditions on labor force participation and attempt to isolate the so-called "discouraged worker" effect,[2] and cross-section studies which devote much attention to wage and income elasticities of labor supply.[3] Both classes of study have their methodological shortcomings,[4] but they have found that supply elasticities are generally low (less than one) and that labor force participation is generally pro-cyclical, rising during expansions and falling during recessions.

---

1. Article surveying the recent literature may be found in Parnes and Cain-Watts, pp. 1-13.
2. See, for examples, Dernberg-Strand; Tella; Bowen-Finegan.
3. See, for examples, Mincer (1962) Cain; Bowen-Finegan; Cain-Watts; Cohen-Rea-Lerman.
4. See Mincer (1968) for a critical review of these studies.

Models designed to forecast future labor force participation have generally been based on the findings of the time-series analyses. One of the more notable of these models is the BLS model associated with Dennis Johnston and Sophia Travis.[1] This model, the basis of supply projections for a large number of manpower forecasting studies,[2] has been criticized for failing to take adequate account of the discouraged worker effect.[3] On the other hand, the critics have also been criticized for producing rates of growth in labor force participation that are unrealistic.[4]

Relative Supply and Mobility

In his rather thorough review of the literature on labor mobility, Parnes concludes that there is a "substantial flexibility" in the United States labor force; but he is unable to attribute this flexibility to purely economic or market factors.[5] A labor market in which relative supply adapts relatively quickly is a labor market in which the short-run wage-elasticity of supply is relatively large. This implies that large shifts in the demand schedule for labor (caused presumably by technological innovations) will be accommodated rapidly on the supply side with relatively small changes in relative wages. A low wage-elasticity of supply would imply that large shifts in the demand for labor (induced by technological innovation) will result in precipitous declines or skyrocketing increases in relative wages. Where wages are inflexible downward, these shifts could produce substantial amounts of unemployment. Thus, the markets in which the

1. Johnston.
2. See, for example, Lecht, BLS (1966, 1970), Wool.
3. Dernberg-Strand-Dukler and Cooper-Johnston.
4. Easterlin.
5. Parnes.

social costs of changing technology could be potentially high are those markets in which wage-elasticities of supply are relatively low. Supply responsiveness to wage changes have been estimated for a number of skills ranging from domestics to college graduates. These estimates have been produced using a variety of techniques and methods. Generally speaking, supply has been measured in either one of two ways: (1) as the number of workers supplying their services in a particular labor market or (2) as the flow of new entrants into this labor market. Examples of the former may be found in the studies of the supply of nurses,[1] domestics,[2] teachers,[3] doctors,[4] dentists,[5] and low-skilled labor.[6] Attempts have also been made in studies of the supply of engineers.[7] Examples of the latter may be found in studies of college-trained manpower,[8] engineers,[9] military manpower,[10] and nurses.[11]

Estimates of short-run supply responsiveness to wage changes have generally been low for such occupations as nurses, doctors, and dentists--occupations which generally have long training periods. Longer run responses, in the form of new entrants, are generally more sensitive to variables that act as proxies for rates of return to investment in training. Some studies note that mobility (particularly geographic mobility) is sometimes hindered by occupational licensing requirements, particularly in the professional

---

1. Benham; Altman (1971).
2. Mattaila; Stigler.
3. Gramm.
4. Benham-Maurizi-Reder.
5. Ibid.
6. Crandall-MacRae-Yap.
7. O'Connell; Black-Stigler.
8. Freeman.
9. Cain-Freeman-Hansen.
10. Altman (1969), Altman-Barro, Altman-Fechter, Oi, Cook, Fechter (1967, 1972), Fisher, Gray, Kim-Farrell-Clague, pp. 79-120.
11. Altman (1971).

occupations.[1] On the other hand, studies of the labor market for public employees have produced rather large supply elasticities.[2]

The evidence on labor supply behavior also reveals that a substantial fraction of the observed variance in this behavior can be attributed to factors other than wages or rates of return. This does not necessarily mean that economic forces are not operating in these markets or that, because of this, adjustments in these markets will be very costly. Rather, it means that a large fraction of labor market adjustment occurs through shifts in both supply and demand curves.

The evidence available from existing studies of labor-force participation and relative supply is not strong enough to be used to conclude that labor supply is highly adaptable. However, it is clearly sufficient to warrant the conclusion that labor markets are not inflexible.

Technological Unemployment: Some Empirical Evidence

An alternative way of viewing adaptability is to evaluate the outcomes arising from shifts in market supply or demand schedules. A crude measure of these outcomes is the amount of unemployment that exists in labor markets. This measure probably understates the impact of technology because it omits displaced workers who have dropped out of the labor force or who are working involuntarily part-time as a result of technological change.

Unemployment is traditionally disaggregated into four analytic categories: cyclical, structural, seasonal, and frictional. The kind of unemployment that is produced by technological change is generally classified as structural;

1. An early statement of this can be found in Friedman-Kuznets. More recent evidence is cited in Holen and Benham-Maurizi-Reder.
2. Ashenfelter.

i.e., it is unemployment that is not the result of normal labor market turnover (frictional), seasonal variations in supply or demand conditions in the labor market (seasonal), or inadequate demand (cyclical). Moreover, within any given analytic category of unemployment, the rate will depend on the amount of turnover in the labor market and the average duration of the unemployment. Frictional unemployment is largely determined by turnover, whereas structural unemployment is largely the result of duration. In an interesting study of unemployment at full employment, Hall concludes that ". . . the evidence on the duration of unemployment and on individuals who are not in the labor force suggests rather strongly that chronic inability to find a job is not a problem found by a significant number of people when the economy is at full employment. The real problem is that many workers have frequent short spells of unemployment."[1] Hall's data suggest that only one-fourth to one-sixth of the unemployed in a relatively full employment period were unemployed for more than 14 weeks. Unemployment averaged roughly 2.8 million in the year of his analysis; therefore, the number of long-term unemployed ranged between approximately 470 and 700 thousand. This would suggest that technological unemployment is a relatively minor phenomenon in a fully employed economy.

---

1. Hall, p. 387.

VI

CONCLUSIONS AND RECOMMENDATIONS

The conclusion that emerges from this review is that our capability to forecast the impact of technology on skill-mix is extremely limited. Major barriers to the development of such capability include:

- an ability to forecast future innovations--both their occurrence and the process of their integration into the production process.
- limitations in the specification of production relationships so that labor can be easily disaggregated into skills.
- inadequate measures of skill.
- an inability to disentangle changes in labor-productivity and skill-mix attributable to changes in technology from changes attributable to other determinants.
- unreasonable assumptions in the forecasting models--specifically, either that skill-mix and labor productivity are constant or that they will change in accordance with past trends.

Overcoming these barriers would be a priority research objective if labor markets were not able to adapt readily to changing technology. A brief review of the literature on adaptability on the supply side of the labor market led to the conclusion that there is a reasonable amount of flexibility, but that it is not perfect.

It is difficult to make recommendations without some knowledge of the options available to the decision maker. However, if rational decisions are to be made about the allocation of resources to improving the existing capacity to forecast the impact of technology on skill utilization, the magnitude and location of so-called "technological" unemployment should first be assessed.

The evidence reviewed in this study suggests that the problem of technological unemployment is not one which would justify extreme measures. However, given the magnitude of the problem, if it is decided that forecasting models designed to pinpoint the future impact of technology would be desirable for policy making, then a rational strategy would encourage future research in the following areas:

1. the determinants of innovation and the factors that affect the length of time it takes to integrate into the production process.
2. development and estimation of the parameters of production functions capable of deriving employment by skill.
3. development and evaluation of alternative skill taxonomies that will aggregate labor into more mutually exclusive, homogeneous units.
4. the determinants of past changes in labor-productivity and skill-mix.
5. analysis of the predictive performance of existing forecasting models.

Some of these areas are currently being investigated--particularly area 1 and labor-productivity in area 4. However, there is little systematic work being done in areas 2, 3, 4 (skill-mix) and 5. These are the areas to which future research resources should be allocated if the existing forecasting capability is to be improved.

While the direct benefits of such research may appear to be small, there are externalities that enhance the prospective value of such a research undertaking. The problem is analytically equivalent to the problem of assessing the impact of any factor that might affect the demand for labor by level of skill. A particularly urgent area of policy interest at the present time is the question of the impact of alternative foreign trade policies on American labor markets. The models discussed in this survey are also applicable to this particular policy issue, but they have the same shortcomings. In addition, the recent passage of the Manpower Revenue Sharing Act provides incentives to develop manpower requirements models that can be applied at the regional (state and local) level. Presumably areas expected to experience large downward shifts in the demand for their labor ought to be provided with more funds than areas that are expected to grow, other things equal. These externalities suggest that all factors (including technology) expected to change manpower skill requirements are going to be important for policy considerations and that this importance is not expected to diminish for quite some time. Taken together, these factors would constitute justification for supporting a general research agenda to improve the existing ability to anticipate changes in manpower skill requirements that will arise from technology and from other important determinants of labor demand.

BIBLIOGRAPHY

Ahamad, Bashir and Mark Blaug, eds. The Practice of Manpower Forecasting (San Francisco: Jossey-Bass, Inc., 1973).

Ahamad, Bashir. A Projection of Manpower Requirements by Occupation in 1975: Canada and its Regions (Ottawa, Department of Manpower and Immigration, 1969.)

Almon, Clopper, Jr. (1967) The American Economy to 1975 (New York: Harper and Row Publishers, 1967).

Almon, Clopper, (1970a) "Investments in Input-Output Models and the Treatment of Secondary Products," in Carter, Anne P., and A. Brody, eds., Applications of Input-Output Analysis, (Amsterdam: Nort-Holland, 1970), pp. 103-116.

Almon, Clopper, (1970b) "The Maryland Interindustry Forecasting Model," paper presented to the Symposium on Models of Economics, Novosibirsk, U.S.S.R., June, 1970.

Alterman, Jack. "An Overview of BLS Projections," Monthly Labor Review, Vol. 96, No. 12 (Dec. 1973), pp. 3-7.

Altman, Stuart H. (1969) "Earnings, Unemployment, and the Supply of Enlisted Volunteers," Journal of Human Resources, Vol. 4, No. 1 (Winter 1969), pp. 38-59.

Altman, Stuart H. (1971) Present and Future Supply of Registered Nurses (Washington, D.C.: U.S. Government Printing Office, November 1971).

Altman, Stuart H. and Robert J. Barro. "Officer Supply: The Impact of Pay, the Draft, and the Vietnam War," American Economic Review, Vol. 61, No. 4 (September 1971), pp. 649-664.

Altman, Stuart H. and Alan E. Fechter. "The Supply of Military Personnel in the Absence of a Draft," American Economic Review, Vol. 57, No. 2 (May 1967), pp. 19-31.

Ashenfelter, Orley. "Demand and Supply Functions for State and Local Employment: Implications for Public Employment Programs," unpublished manuscript, no date.

Bain, Trevor. Labor Market Experience for Engineers during Periods of Changing Demand, (New York: Center for Policy Research, Inc., November, 1973).

Benham, Lee. "The Labor Market for Registered Nurses: A Three-Equation Model," Review of Economics and Statistics, Vol. 52, No. 3 (August 1971), pp. 246-252.

Benham, L., A. Maurizi, and M.W. Reder. "Migration, Location, and Remuneration of Medical Personnel, Physicians and Dentists," Review of Economics and Statistics, Vol. 50, No. 3 (August 1968), pp. 332-347.

Bezdek, Roger H. (1972) "Alternative Manpower Forecasts to 1975 and 1980: Second Guessing the U.S. Department of Labor," Proceedings of Annual Meeting of the American Statistical Association, 1972.

Bezdek, Roger H. (1973) "Models and Methods for Long-Range Economic and Manpower Forecasting," paper presented at the annual meeting of the Operations Research Society of America, Nov., 1973.

Bezdek, Roger H., and Barry Getzel. "Alternative Forecasts of the Job Content and Skill Requirements of the U.S. Economy in 1980," Technological Forecasting and Social Change, Vol. 5, No. 2 (1973), pp. 205-214.

Bezdek, Roger H., and Bruce Hannon. "Energy and Manpower Effects of Alternative Uses of the Highway Trust Fund," Center for Advanced Computation, University of Illinois, 1973.

Blank, D.M. and George J. Stigler. The Demand and Supply of Scientific Personnel (New York: National Bureau of Economic Research, 1957).

Bowen, William G. and T. Aldrich Finegan. The Economics of Labor Force Participation (Princeton: Princeton University Press, 1969).

Bowles, Samuel S. "Aggregation of Labor Inputs in the Economics of Growth and Planning: Experiments with a Two-Level C.E.S. Function," Journal of Political Economy, Vol. 78, No. 1 (January/February 1970), pp. 68-81.

Bright, James R. "The Relationship of Increasing Automation and Skill Requirements," in National Commission on Technology, Automation, and Economic Progress, Technology and the American Economy, Appendix Vol. II, The Employment Impact of Technological Change, pp. 203-222.

Bureau of Labor Statistics (1966a). "America's Industrial and Occupational Manpower Requirements, 1964-1975," In National Commission on Technology, Automation, and Economic Progress, Technology and the American Economy, Appendix Vol. I, The Outlook for Technological Change and Employment (Washington, D.C.: National Commission on Technology, Automation, and Economic Progress, February 1966), pp. 1-202.

Bureau of Labor Statistics (1966b). Projections 1970: Interindustry Relationships, Potential Demand, Employment, Bulletin 1566 (1966).

Bureau of Labor Statistics (1970). Patterns of U.S. Economic Growth, Bulletin 1672 (1970).

Bureau of Labor Statistics (1972a). Tomorrow's Manpower Needs: Vol. IV, The National Industry-Occupational Matrix and Other Manpower Data, Bulletin 1737 (1972) Revised 1971.

Bureau of Labor Statistics (1972b). Projections of the Post-Vietnam Economy, 1975, Bulletin 1733 (1972).

Cain, Glen G. Married Women in the Labor Force: An Economic Analysis (Chicago: The University of Chicago Press, 1966).

Cain, Glen G., Richard B. Freeman, and W. Lee Hansen. Labor Market Analysis of Engineers and Technical Workers (Baltimore: The Johns Hopkins University Press, 1973).

Cain, Glen G. and W. Lee Hansen. "An Economic Approach to Classification of Occupations," Monthly Labor Review, Vol. 90, No. 2 (February 1967), pp. 48-52.

Cain, Glen G. and Harold W. Watts, eds. Income Maintenance and Labor Supply (Chicago: Markham Books, 1973).

Cartter, Allan M. "The Supply and Demand for College Teachers," Journal of Human Resources, Vol. 1, No. 1 (Summer, 1966), pp. 1-38.

Cook, A.A., Jr. "Quality Adjustment and the Excess Supply of Air Force Volunteers," Review of Economics and Statistics, Vol. 54, No. 2 (May 1972), pp. 166-171.

Cooper, Sophia and Denis F. Johnston. "Comments on the Dernberg-Strand-Dukler Approach," Industrial Relations, Vol. 6, No. 1 (October 1966), pp. 69-78.

Crandall, R. W., C.D. MacRae and L.Y.L. Yap. An Economic Model of the Low-Skill Labor Market, Urban Institute Working Paper 3603-01 (Washington, D.C.: Urban Institute, August 1973).

Christ, Carl. "A Review on Input-Output Analysis" in Input-Output Analysis: An Appraisal, National Bureau of Economic Research Studies in Income and Wealth, Volume 18 (Princeton: Princeton University Press, 1955), pp. 137-169.

Christensen, L.R., D.W. Jorgenson, and L.J. Lau. "Transcendental Logarithmic Production Frontiers," Review of Economics and Statistics, Vol. 55, No. 1 (February 1973), pp. 28-45.

Cohen, Malcom S., Samuel A. Rea, Jr., and Robert J. Lerman. A Micro Model of Labor Supply, Bureau of Labor Statistics Staff Paper No. 4 (Washington, D.C.: U. S. Government Printing Office, 1970).

Darmstadter, Joel. "Manpower in a Long-Term Projection Model," Industrial Relations, Vol. 5, No. 3 (May 1966), pp. 28-58.

Department of Education and Science. The Demand for Supply of Teachers, 1963-1986, Ninth Report of the National Advisory Council on the Training and Supply of Teachers (London: HMSO, 1965).

Dernberg, Thomas and Kenneth Strand. "Hidden Unemployment, 1953-1962: A Quantitative Analysis by Age and Sex," American Economic Review, Vol. 56, No. 1 (March 1966), pp. 71-95.

Dernberg, Thomas, Kenneth Strand, and Judith Dukler. "A Parametric Approach to Labor Force Projections," Industrial Relations, Vol. 6, No. 1 (October 1966), pp. 46-68.

District of Columbia Manpower Administration. Manpower Directions in Metropolitan Washington, D.C.: A Forecast of Manpower Demands and Employment by Industry and Occupation, 1972-1980 (Washington, D.C.: 1972).

Dougherty, C.R.S. "Estimates of Labor Aggregation Functions," Journal of Political Economy, Vol. 80, No. 6 (November/December 1972), pp. 1101-1119.

Durand, John D. The Labor Force in the United States, 1890-1960 (New York Social Science Research Council, 1948).

Engineering Manpower Commission. Demand for Engineers and Technicians (New York: Engineers Joint Council, 1968).

Fabricant, S. Basic Facts on Productivity Change (New York: Columbia University Press, 1959).

Fair, Ray C. Short-Run Demand for Workers and Hours (Amsterdam: Nort-Holland, 1969).

Fechter, Alan E. (1972) Army Enlistments and the All-Volunteer Force: The Application of an Econometric Model, Institute for Defense Analysis Paper P-845 (Washington, D.C.: Institute for Defense Analysis, 1972).

Fechter, Alan E. (1967) The Supply of First-Term Military Officers, Institute for Defense Analysis Study S-290 (Washington, D.C.: Institute for Defense Analysis, 1967).

Ferber, Robert and Kyshei Sasaki. "Labor Force and Wage Projections in Hawaii," Industrial Relations, Vol. 5, No. 3 (May 1966), pp. 72-85.

Fisher, A.C. "The Costs of the Draft and the Costs of Ending the Draft," American Economic Review, Vol. 59, No. 3 (June 1969), pp. 239-255.

Fisher, F.M., L.R. Klein, and Y. Shinkai. "Price and Output Aggregation in the Brookings Econometric Model" in Duesenberry, J.S., Gary Fromm, and Edwin Kuh, eds., The Brookings Quarterly Econometric Model of the United States, (Chicago: Rand McNally and Co., 1965), pp. 653-679.

Fortune, J. Neill and Peter Ross. Supply and Requirement Projections of Engineers--Canada, 1976 (Ottawa: Department of Manpower and Immigration).

Freeman, Richard B. The Market for College-Trained Manpower (Cambridge: Harvard University Press, 1971).

Friedman, Milton. Essays in Positive Economics (Chicago: University of Chicago Press, 1953), pp. 3-43.

Friedman, Milton and Simon Kuznets. Income from Independent Professional Practice (New York: National Bureau of Economic Research, 1954).

Fromm, Gary. "Forecasting, Policy Simulation and Structural Analysis: Some Comparative Results of Alternative Models," in American Statistical Association Business and Economic Statistics Section, Proceedings, 1964, pp. 6-17.

Fromm, Gary and L.R. Klein. "The Brookings Model Volume: A Review Article: A Comment," The Review of Economics and Statistics, Vol. 50, No. 2, (May, 1968), pp. 235-240.

Gramlich, E.M. "The Demand for Different Skill Classes of Labor," Unpublished manuscript, May, 1972.

Gramm, Wendy Lee. "The Labor Force Decision of Married Female Teachers: A Discrimination Analysis Approach," Review of Economics and Statistics, Vol. 55, No. 3 (August 1973), pp. 341-348.

Gray, B.C. "Supply of First-Term Military Enlistees," in Studies Prepared for the President's Commission on an All-Volunteer Armed Force, Vol. 1 (Washington, D.C.: U.S. Government Printing Office, 1970).

Haase, Peter E. "Technological Change and Manpower Forecasts," Industrial Relations, Vol. 5, No. 3 (May 1966), pp. 59-71.

Hall, Robert E. "Why is the Unemployment Rate So High at Full Employment?" Brookings Papers on Economic Activity, 3:1970, pp. 369-410.

Hansen, W. Lee. "Labor Force and Occupational Projections," Proceedings of the Eighteenth Annual Winter Meeting of the Industrial Relations Research Association, pp. 10-20.

Holen, Arlene. "Effects of Professional Licensing Arrangements on Interstate Labor Mobility and Resource Allocation," Journal of Political Economy, Vol. 73 (October 1965), pp. 492-498.

Horowitz, Morris A. and Irwin L. Herrnstadt. "Changes in the Skill Requirements of Occupations in Selected Industries," in National Commission on Technology, Automation, and Economic Progress, Technology and the American Economy, Appendix Vol. II, The Employment Impact of Technological Change, pp. 223-288.

Hollister, R.G. "The Economics of Manpower Forecasting," International Labour Review, Vol. 89, No. 4 (April 1964), pp. 371-397.

Johnston, Denis F. "Population and Labor Force Projections," Monthly Labor Review, Vol. 96, No. 12 (December 1973), pp. 8-17.

Kendrick, John W. Productivity Trends in the United States (Princeton: University Press, 1962).

Kennedy, Charles and A. P. Thirlwall. "Surveys in Applied Economics: Technical Progress," The Economic Journal, Vol. 82, No. 325 (March 1972), pp. 11-72.

Kim, K.H., S. Farrell, and E. Clague. The All-Volunteer Army: An Analysis of Demand and Supply (New York: Praeger Publishers, 1971)

Klein, Lawrence R. An Essay on the Theory of Prediction (Chicago: Markham Publishing Company, 1970).

Klotz, Ben. "A Simple Proposal to Improve Forecasting Models," Urban Institute, November 1973.

Koopman, Tjalling C. Three Essays on the State of Economic Science (New York: McGraw-Hill Publishers, 1957).

Kresge, David T. "Price and Output Conversion: A Modified Approach," in Duesenberry, James S., Gary Fromm, Lawrence R. Klein, and Edwin Kuh, eds., The Brookings Model: Some Further Results. (Chicago: Rand McNally and Co., 1969), pp. 85-109.

Lecht, Leonard. Manpower Needs for National Goals in the 1970's (New York: Praeger Publishers, 1969).

Long, Clarence. The Labor Force Under Changing Income and Employment (Princeton: Princeton University Press, 1958).

Massell, Benton F. "A Disaggregated View of Technical Change," Journal of Political Economy, Vol. 68, No. 6 (December 1961), pp. 547-557.

Mattila, J. Peter. "The Effect of Extending Minimum Wages to Cover Household Maids," Journal of Human Resources, Vol. 8, No. 3 (Summer 1973), pp. 365-382.

McCormick, William W. and Charles M. Franks. "A Method for Forecasting Regional Industry and Occupational Employment Levels," Western Economic Journal, Vol 5 (June 1967), pp. 233-256.

McCracken, M.C. An Overview of Candide Model 1.0 (Ottawa, Ontario: Economic Council of Canada, February 1973).

McGilvray, James and David Simpson. "Some Tests of Stability in Interindustry Coefficients," Econometrica, Vol. 37, No. 2 (April 1969), pp. 204-221.

Mincer, Jacob. "Labor Force Participation of Married Women: A Study of Labor Supply," in National Bureau of Economic Research, Aspects of Labor Economics (Princeton: Princeton University Press, 1962), pp. 63-97.

Mincer, Jacob and Victor Zarnowitz. "The Evaluation of Economic Forecasts," in Mincer, J., ed. Economic Forecasts and Expectations: Analyses of Forecasting Behavior and Performance (New York: Columbia University Press, 1969).

National Commission on Technology, Automation, and Economic Progress. Technology and the American Economy (Washington, D.C.: U.S. Government Printing Office, 1966).

National Education Association, Research Division. Teacher Supply and Demand in Public Schools, 1964, Research Report 1964-R9 (Washington, D.C.: National Education Association, 1964).

National Planning Association. Projections of Economic and Demographic Variables in New York State Development Regions (Washington, D.C.: National Planning Association, 1966).

National Science Foundation. Engineers and Technicians in the 1960's: Requirements and Supply (Washington, D.C.: National Science Foundation, 1963).

Nerlove, M. "Recent Empirical Studies of the C.E.S. and Related Production Functions," in M. Brown, ed. The Theory and Empirical Analysis of Production (New York: Columbia University Press, 1967).

O'Connell, John F. "The Labor Market for Engineers: An Alternative Methodology," Journal of Human Resources, Vol. 7, No. 1 (Winter 1972), pp. 71-86.

Oi, Walter Y. "The Economic Cost of the Draft," American Economic Review, Vol. 57, No. 2 (May 1967), pp. 39-62.

Parnes, Herbert S. "Labor Force and Labor Markets," in G.G. Somers, ed. A Review of Industrial Relations Research (Madison: Industrial Relations Research Association, 1970), pp. 1-78.

Psacharopoulos, G. and K. Hinchliffe. "Further Evidence on the Elasticity of Substitution Between Different Types of Educated Labor," *Journal of Political Economy*, Vol. 80, No. 4 (July/August 1972), pp. 786-792.

Rosenthal, Neal H. "Projected Changes in Occupations," *Monthly Labor Review,* Vol. 96, No. 12 (December 1973), pp. 18-26.

Ross, Peter. *Supply and Requirement Projections of Physicists, Canada, 1976* (Ottawa: Department of Manpower and Immigration, 1971).

Scott, Graham C. *U.S. Economic and Demographic Projections: 1972-81* (Washington, D.C.: National Planning Association, 1973).

Scoville, James G. *The Job Content of the U.S. Economy, 1940-1970* (New York: McGraw-Hill, Inc., 1969).

Solow, Robert. *The Nature and Sources of Unemployment in the United States* (Stockholm: 1964).

Spellman, William E. *Projections of Occupational Requirements for Kansas in 1980* (Manhattan, Kansas: Department of Economics, Kansas State University, 1970).

Stekler, H.O. *Economic Forecasting* (New York: Praeger Publishers, 1970).

Stigler, George. *Domestic Servants in the United States,* Occasional Paper 24 (New York: National Bureau of Economic Research, April 1946).

Striner, Herbert. "An Evaluation of the Report of the National Commission on Technology, Automation, and Economic Progress," *Proceedings of the Nineteenth Annual Winter Meeting of the Industrial Relations Research Association.*

Sundarajan, V., *A Flexible Coefficient Bridge Model: Trend-Cycle Adjustments in Input-Output Analysis*, Unpublished doctoral dissertation, Harvard University, 1970.

Swerdloff, Sol. "Manpower Projections: Some Conceptual Problems and Research Needs," *Monthly Labor Review*, Vol. 89, No. 2 (February 1966).

Tella, Alfred. "Labor Force Sensitivity to Employment by Age, Sex," *Industrial Relations*, Vol. 4, No. 2 (February 1965), pp. 69-83.

Theil, Henri. *Applied Economic Forecasting* (Amsterdam: Nort-Holland, 1966).

Theil, Henri. *Optimal Decision Rules for Government and Industry* (Amsterdam: Nort-Holland, 1964).

Watson, Donald A. *An Input-Output Model for State Manpower Projections* (Eugene, Oregon: Bureau of Business and Economic Research, University of Oregon, 1970).

Wool, Harold. The Labor Supply for Lower Level Occupations: Interim Report (Washington, D.C.: National Planning Association, 1973).

Woytinsky, W.S. Three Aspects of Labor Dynamics (Washington, D.C.: Social Science Research Council, 1940).